'Safe' hot water and surface temperatures

Health Guidance Note

London: The Stationery Office

NATIONAL HEALTH SERVICE IN SCOTLAND
ESTATES ENVIRONMENT FORUM

NHS Estates
An Executive Agency of the Department of Health

First published 1992
New edition 1998

ISBN 0-11-322158-4

Standing order service

Are you making full use of The Stationery Office's Standing Order Service?

The Standing Order Service is a free monitoring of the publications of your choice from over 4000 classifications in 30 major subject areas. We send you your books as they are published, along with an invoice.

With a standing order for class 05.03.037 you can be supplied automatically with further titles in this series as they are published.

The benefits to you are:

- automatic supply of your choice of classification on publication;
- no need for time-consuming and costly research, telephone calls and scanning of daily publication lists;
- saving on the need and the costs of placing individual orders.

We can supply a wide range of publications on standing order, from individual annual publkications to all publications on a selected subject. If you do not already use this free service, or think you are not using it to its full capacity, why not contact us and discuss your requirements?

You can contact us at:

The Stationery Office
Standing Order Department
PO Box 276
London SW8 5DT

Tel: 0171-873 8466; fax 0171-873 8222

We look forward to hearing from you.

About this series

Health Guidance Notes are an occasional series of publications prepared by NHS Estates. They respond to changes in Departmental policy or reflect changing NHS operational management and cover specific topics.

The Health Guidance Notes are intended to appraise general managers and chief executives of these changes and to offer recommendations on appropriate action.

The contents of this Health Guidance Note are endorsed by NHS Management Executive for the NHS in England, Welsh Health Estates for the NHS in Wales, Health Estates for the NHS in Northern Ireland, and the Healthcare Engineering and Environment Unit for the NHS in Scotland.

Executive summary

This Health Guidance Note recommends how to meet employers' legal duty of care in respect of the risk of scalding and burning from hot water and hot surfaces. It applies in all healthcare premises, and those premises registered under the Registered Homes Act 1984. It may also be appropriate in non-registered premises, for example sheltered accommodation, where the occupants are equally at risk.

This HGN has been revised to take account of new guidance published in HTM 2027 – 'Hot and cold water supply, storage and mains services' and HTM 2040 – 'The control of legionellae in healthcare premises: a code of practice', and the introduction of Model Engineering Specification D08 – 'Thermostatic mixing valves (healthcare premises)'.

A risk assessment should be carried out to identify potential risks and the action to be taken. A planned programme of upgrade should be undertaken, with appropriate interim precautions taken in areas shown to be at risk.

Hot water temperatures

The guidance gives maximum set hot water temperatures for a range of applications, for example 44°C for bath fills and 41°C for showers. Higher hot water temperatures may only be used following a thorough risk assessment. The guidance recommends that where patients are considered to be vulnerable to scalding, thermostatic mixing valves (TMVs) should be used to control hot water temperatures.

There are three types of mixing valve:

Type 1 – a mechanical mixing valve;

Type 2 – a thermostatic mixing valve;

Type 3 – a thermostatic mixing valve with enhanced performance.

Type 3 TMVs are recommended in areas of special risk, that is, where patients are immersed in water or where skin is exposed to running hot water. All Type 3 TMVs must have undergone third-party testing to the requirements of Model Engineering Specification D08 and been awarded a certificate under 'The TMV Scheme' managed by WRc. All valves should be subject to a range of commissioning and in-service tests, which are described in Chapter 5 of this HGN.

Hot surface temperatures

The surface temperature of space heating devices such as storage heaters and radiators should be controlled to prevent patients being burned. This guidance recommends that surface temperatures should not exceed 43°C.

Contents

1.0 Application

General

HS(G)104:1993 – Health and Safety in Residential Care Homes; IND(G)253L-6/97-C150 – Controlling legionella in nursing homes and residential care homes.

Northern Ireland: Nursing Homes and Nursing Agencies Act (Northern Ireland) 1971; Scotland: Registered Establishments (Scotland) Act 1987.

1.1 This guidance recommends how to meet employers' legal duty of care in respect of the risk of scalding and burning from hot water and hot surfaces. It applies in all healthcare premises, and those premises registered under the Registered Homes Act 1984. It may also be appropriate in non-registered premises, for example sheltered accommodation, where the occupants are equally at risk.

1.2 All patients, residents, visitors and staff must be presumed to be potentially at risk, but some are more vulnerable to scalding and burning than others; Figure 1 illustrates the time and temperature relationships which result in partial and full thickness burns.

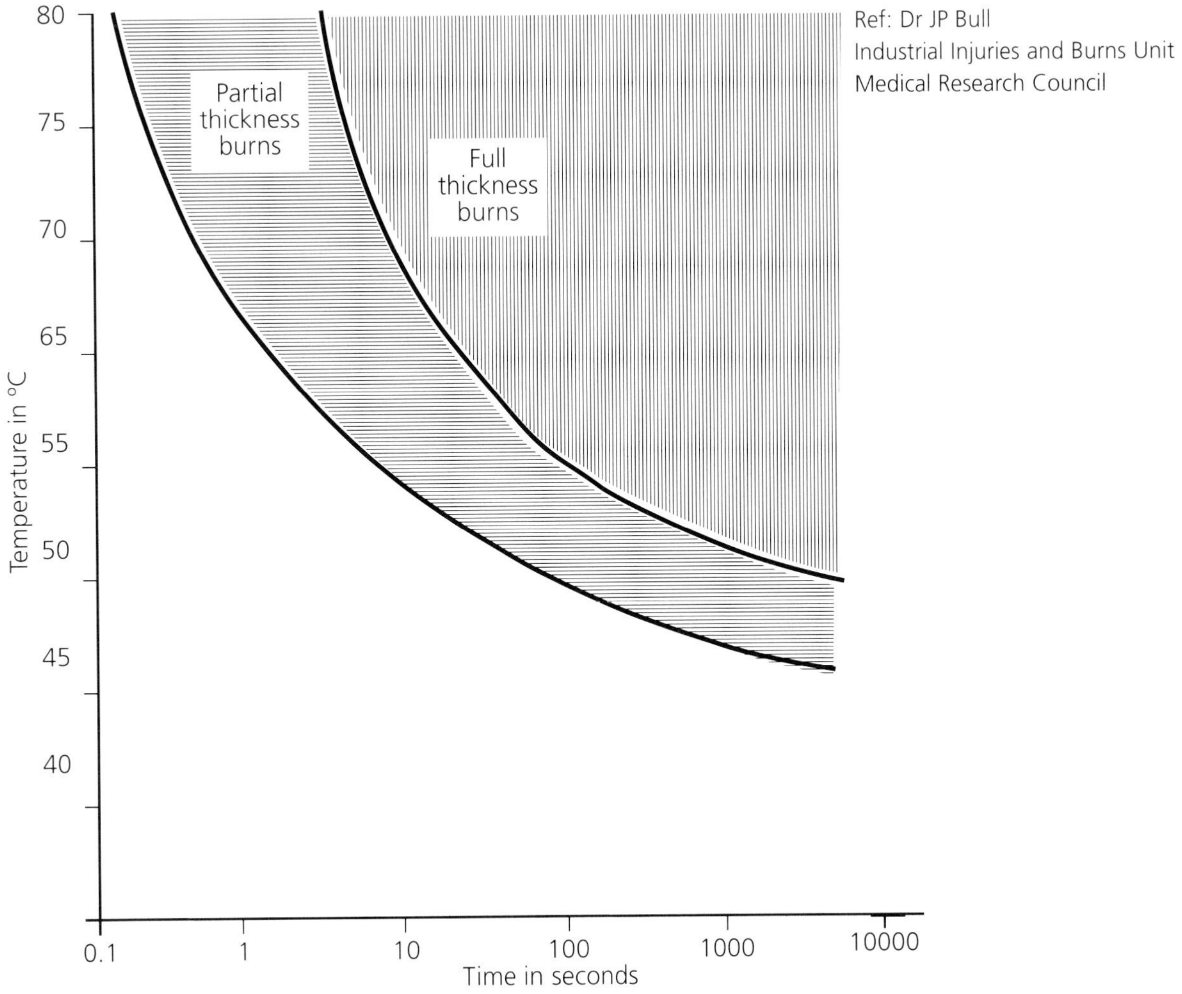

Figure 1 Temperature and duration of exposure sufficient to cause burns in thin areas of skin

Rationale for producing new guidance

1.3 New guidance published in HTM 2027 – 'Hot and cold water supply, storage and mains services' and HTM 2040 – 'The control of legionellae in healthcare premises: a code of practice', together with the modern advances in the technology of mixer valve control, prompted a review of the guidance contained in the last edition of this HGN, published in 1992.

1.4 The review concluded that risk of injury from scalding and methods of controlling that risk should be re-assessed, and that in areas where patients are considered to be vulnerable to scalding, thermostatic mixing valves (TMVs) should be used to control hot water temperatures (see Table 1).

1.5 Following consultation with the NHS, manufacturers of TMVs, factor houses and WRc Evaluation and Testing Centre Ltd (WRc), a new Model Specification (D08) for TMVs for higher risk applications has been issued, and is operational as from 2 April 1998.

1.6 'D08 Thermostatic mixing valves (healthcare premises)' describes different types of TMVs capable of operating over a range of temperatures and pressures for washing facilities, for example baths, showers, bidets and washhand basins. The decision whether to use a D08 type TMV should be made following a risk assessment.

1.7 In the process of preparing D08, NHS Estates and manufacturers of TMVs have also co-operated in establishing a Third Party Certification Scheme which is run by WRc.

1.8 Further advice on the application of this type of valve and the Third Party Certification Scheme is given in paragraphs 3.8 to 3.11.

Safe surface temperatures

1.9 Safe surface temperatures are also dealt with in this HGN.

2.0 Action

Health and Safety at Work (Northern Ireland) Order

The Management of Health and Safety at Work Regulations (Northern Ireland)

2.1 Employers have duties under the Health and Safety at Work etc Act 1974 to ensure, so far as is reasonably practicable, the health and safety of their patients, residents, visitors and staff. The Management of Health and Safety at Work Regulations 1992 require employers to carry out assessments of risk and to implement the findings of assessments. Chief executives may delegate tasks to managers to fulfil their duties.

2.2 The recommendations regarding water and surface temperatures apply to all ward accommodation, residents' rooms and those areas to which patients, residents and visitors have free access (including public areas). Where the patient or resident is under adequate continuous supervision, for example diagnostic and treatment rooms, the use of low surface temperature radiators may not be warranted.

A record of the actions taken will also be needed.

2.3 Risk assessment must be carried out to determine:

a. the need to take action;

b. the action to be taken;

c. the consequences thereof.

2.4 When a risk assessment is carried out on existing premises and the risk to the user is considered higher than acceptable, a planned programme of upgrading should be undertaken, with appropriate interim precautions taken in areas shown to be at risk.

2.5 Until the recommended precautions are put into effect, staff should be made aware of the potential danger and take the necessary steps to protect patients, residents and visitors. Areas which do not meet these recommendations should be identified, and plans to comply as soon as reasonably practicable should be devised.

2.6 All staff should be familiar with the local procedures and should be encouraged to report circumstances or changes in circumstances which may increase the risk of scalding.

3.0 'Safe' hot water temperature

Introduction

3.1 The hot water distribution temperatures which are required for the control and prevention of legionellae can lead to discharge temperatures in excess of 50°C.

3.2 To prevent injury from scalding, action will be needed to limit water discharge temperatures.

3.3 The severity of scalding depends upon the temperature of the water and the length of time the skin is exposed to it. The maximum set hot water temperatures for outlets accessible to patients, residents, visitors and staff are given in Table 1.

Table 1 Maximum set hot water temperatures

Application	Maximum set hot water temperature °C
Bidet	38
Shower	41
Washbasin	41[1]
Bath	44
Bath	46[2]

[1] For washbasins, washing under running water is assumed.
[2] Bath fill temperatures of more than 44°C should only be considered in exceptional circumstances where there are particular difficulties in achieving an adequate bathing temperature. If a temperature of more than 44°C is to be used then a safe means of preventing access to the hot water should be devised to protect vulnerable patients.

3.4 Prior to this guidance note, bath/shower mixers were set to 43°C. The guidance in this revision specifies a maximum 44°C set point for unassisted bathing and a 41°C set point for showering. For bath/shower mixers fitted in accordance with previous notes (ie, the same temperature to both bath and shower) a risk assessment will be needed to decide which of the following options is to be adopted:

a. a suitable dual temperature product which automatically provides blended water at either 41°C for the shower outlet or 44°C for the bath outlet;

b. disabling of the showering facility;

c. lower bath temperature of 41°C;

d. replacing the existing valve with individual Type 3 products (see paragraphs 3.8 to 3.11).

3.5 Any decision not to control hot water temperatures in kitchens and rehabilitation areas, or where the hot water is used for processes, for example in dishwashers, should be based on an individual risk assessment. In these instances all users should be made aware of the dangers through training.

3.6 All outlets which are not temperature limited should be clearly labelled "very hot water" by means of fixed notices.

3.7 Lower hot water outlet temperatures may be adopted if functionally appropriate. Individual perception of comfort is variable, particularly when showering.

Types of mixing valve

A Third Party Certification Scheme exists only for Type 3 mixing valves (see paragraph 3.11).

3.8 Mixing valves can be used to reduce hot water temperatures to a safe level. There are three types of mixing valve:

Type 1 – a mechanical mixing valve, or tap including those complying with BS 1415: Part 1, or BS 5779 with maximum temperature stop where appropriate.

Type 2 – a thermostatic mixing valve, generally complying with the previous issue of Health Guidance Note issued in 1992 and BS 1415: Part 2, with maximum temperature stop.

Type 3 – a thermostatic mixing valve with enhanced thermal performance complying with NHS Estates Model Engineering Specification D08.

The choice of mixing valve Type 1, 2 or 3 will depend upon risk assessment.

Table 2 Recommended application for mixing valves

Application	Recommended type of mixing valve
Bidet	Type 3
Shower	Type 3
Washbasins staff areas	Type 1
Washbasins for geriatric, paediatric, mentally handicapped, physically handicapped patients or other patients considered to be at risk	Type 3
Washbasins in patient and visitor areas	Type 1 with maximum temperature stop
Bath	Type 3

Type 3 thermostatic mixing valves

3.9 Where there is whole body immersion or skin exposed to running hot water, patients are considered to be at special risk and a Type 3 thermostatic mixing valve should be used to control the temperature.

3.10 When there are changes in hot water and cold water supply pressures or temperatures to the TMV, or when an outlet is turned on after it has been turned off for some time, the temperature of the hot water leaving the outlet will deviate. The Model Specification for Type 3 valves strictly limits the deviation of mixed water temperature under adverse conditions. A diagrammatic expression of the maximum "safe" temperature including these deviations is shown in Figure 2.

3.11 TMVs tested to and meeting the Model Specification requirements will receive a compliance certificate. This will be managed by WRc under the title 'The TMV Scheme'. A list of valves meeting these requirements will be regularly updated in the 'Water Fittings and Materials Directory' produced by WRc every six months and this will be available separately on request from the WRc (see Appendix II for address and contact numbers).

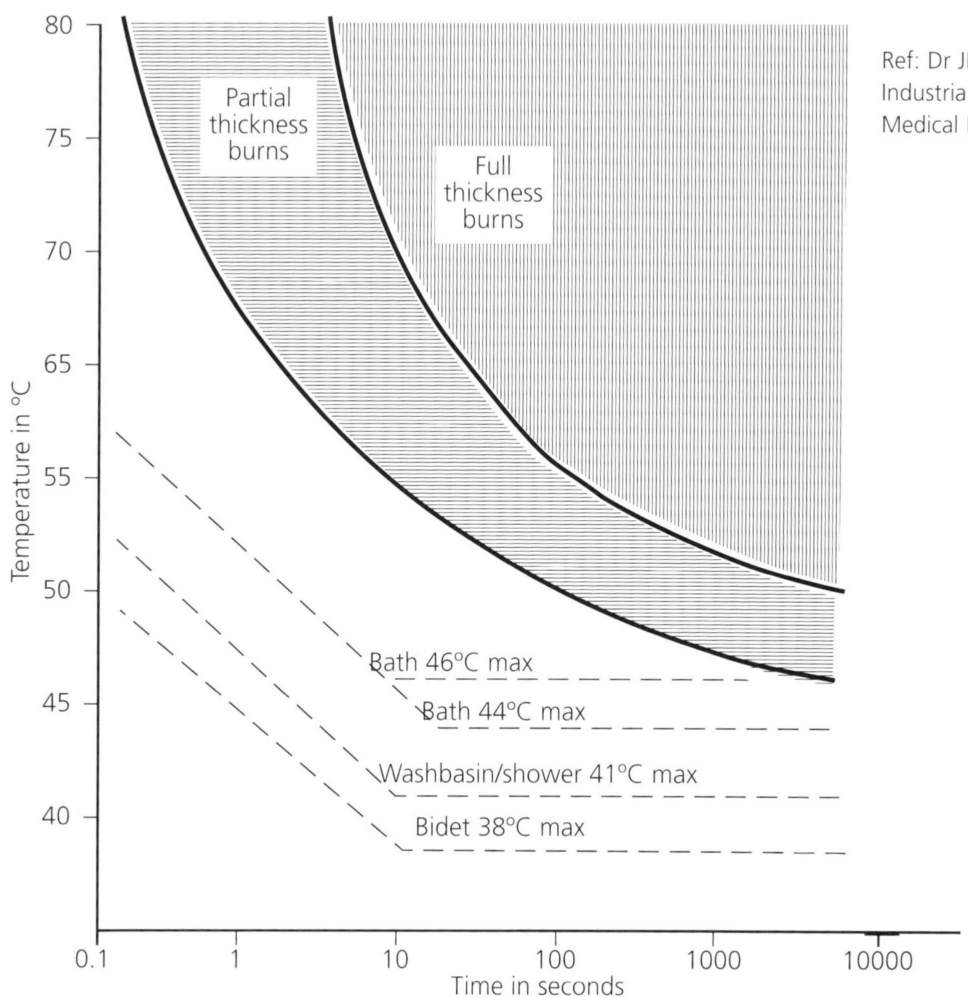

Ref: Dr JP Bull
Industrial Injuries and Burns Unit
Medical Research Council

Figure 2 A diagrammatic expression of the relationship between maximum transient temperature and delivery temperature

4.0 Pre-installation guidance

4.1 In order to select the correct valve for each application, it is essential to identify and record the characteristics of the hot and cold water distribution system that will serve the valve, especially during busy and quiet periods.

4.2 Before selecting a Type 3 valve, the specifier should ensure that the hot and cold water operating pressure is within the limits specified in Model Engineering Specification D08. Before selecting Type 1 and 2 valves, the manufacturer should be consulted to ensure that their valve operates within the hot and cold water pressure of that system. It should be noted that valves may differ in their ability to control temperatures under adverse installation conditions.

4.3 Because mixing valves are vulnerable to debris in the water supply, some protection is required in the form of an integral filter in the product and/or in-line strainers upstream of the valve. TMVs, strainers and isolating valves must be easily accessible for routine maintenance but should be protected from interference by unauthorised personnel.

5.0 Commissioning and in-service tests

5.1 It is important to note that the series of stringent tests designated in Model Specification D08 are only suitable for testing under strict laboratory conditions upon a purpose-built test rig of the correct dimensions. They are not suitable for use as either commissioning or in-service tests.

5.2 The following are recommended test and acceptance criteria for commissioning and in-service tests of Type 3 valves. Some of these tests can also be applied to Type 1 and Type 2 valves.

Pre-commissioning

5.3 To comply with Water Supply Byelaw 55 and BS 6700, all pipework should be flushed after installation, renewal or repair. During flushing, measures should be taken to prevent dirt or debris entering the TMV. Flushing kits supplied by the TMV manufacturers, or other suitable means of safely flushing the system, should be used.

Commissioning

5.4 At the commissioning stage it is recommended that some simple checks and tests on each mixing valve are carried out. As a first step, check:

a. that the designation of the thermostatic mixing valve matches the intended application;

b. that the supply pressures are within the range of operating pressures for the designation of the valve;

c. that the supply temperatures are within the range permitted for the valve and by guidance information on the prevention of legionella etc.

The temperature of the mixed water should be adjusted in accordance with the manufacturer's instructions and the requirement of the application.

5.5 The following tests should then be carried out to establish a performance reference point for future in-service tests:

a. record the temperature of the hot and cold water supplies;

b. record the temperature of the mixed water at the largest draw-off flow rate;

c. record the temperature of the mixed water at a smaller draw-off flow rate appropriate to the application and within the limit set by the manufacturers;

d. isolate the cold water supply to the mixing valve and monitor the mixed water temperature;

e. record the maximum temperature achieved as a result of (d) and the final temperature;

NOTE: The final mixed water temperature must not exceed the values in Table 3.

f. record the equipment, thermometer etc used for the measurements, including make, model, serial number and date of last calibration (the equipment should be the subject of regular certificated calibration).

Table 3 Maximum permissible recorded temperature during site tests of a cold water failure

Application	Mixed water temperature °C
Bidet	40
Shower	43
Washbasin	43
Bath (44°C)	46
Bath (46°C)	48

In-service tests

5.6 The purpose of in-service tests is to regularly monitor and record the performance of the thermostatic mixing valve. Recorded deterioration in performance can indicate the need for service work on the TMV and/or water supply systems and assist in establishing the time interval between tests and service inspections.

5.7 Carry out the procedure 5.5(a) to (c) using the same measuring equipment, or equipment to the same specifications as that used to commission and set the TMV outlet temperature.

5.8 If the mixed water temperature has changed significantly from the previous test results (eg >1°C), record the change and, before re-adjusting the mixed water temperature, check that:

a. all in-line or integral strainers are clean;

b. all in-line or integral check valves or other anti-backsiphonage devices are in good working order;

c. all isolating valves are fully open.

5.9 With an acceptable mixed water temperature, complete the procedure 5.5(a) to (f).

In-service tests should be carried out at a frequency which identifies the need for service work before an unsafe water temperature can result. In the absence of any other instruction or guidance, the procedure described below may be used.

5.10 If at step 5.5(e) the final mixed water temperature is greater than the values in Table 3 and/or the maximum temperature exceeds the corresponding value from the previous test results by more than about 2°C, the need for service work is indicated.

Frequency of in-service tests

5.11 In the absence of any other instruction or guidance on the means of determining the appropriate frequency of in-service testing, the following procedure may be used.

5.12 Six to eight weeks after commissioning, carry out the tests given in paragraph 5.5.

5.13 Twelve to 15 weeks after commissioning, carry out the tests given in paragraph 5.5.

5.14 Depending on the results of 5.12 and 5.13, several possibilities exist:

a. if no significant changes (eg <1°C) in mixed water temperatures are recorded between commissioning and 5.12, or between commissioning and 5.13, the next in-service tests can be deferred to 24 to 28 weeks after commissioning;

b. if small changes (eg 1 to 2°C) in mixed water temperatures are recorded in only one of these periods, necessitating adjustment of the mixed water temperature, the next in-service test can be deferred to 24 to 28 weeks after commissioning;

c. if small changes (eg 1 to 2°C) in mixed water temperatures are recorded in both of these periods, necessitating adjustment of the mixed water temperatures, the next in-service test should be carried out at 18 to 21 weeks after commissioning;

d. if significant changes (eg >2°C) in mixed water temperatures are recorded in either of these periods, necessitating service work, the next in-service test should be carried out at 18 to 21 weeks after commissioning.

5.15 The general principle to be observed after the first two or three in-service tests is that the intervals of future tests should be set to those which previous tests have shown can be achieved with no more than a small change in mixed water temperature. The quality and type of water, local operational standards and knowledge of the hot water systems will also influence the service and test intervals.

6.0 Maintenance

6.1 Maintenance of all types of mixing valve is essential. Scale or small pieces of dirt or debris may restrict the operation of the temperature control mechanism and prevent the valve from cutting off the hot water supply in the event of cold water supply failure. Cleaning the internal working parts, therefore, is an essential part of maintenance. The frequency at which maintenance is carried out will depend upon:

a. results of in-service testing;

b. quality of the water passing through the valve.

6.2 The manufacturer should supply detailed instructions for installation, commissioning, maintenance and temperature adjustment for each type of TMV. The manufacturer's instructions should be strictly followed.

6.3 Maintenance records and the record of the temperatures at the hot water outlet before and after servicing or any alteration to the TMV should be kept (a sample maintenance log sheet is contained in Appendix I). These records should be inspected by the designated and/or responsible officer on a regular basis to ensure standards are being maintained as required. Maintenance and servicing of the TMVs should be provided by appropriately trained staff.

7.0 Certification

7.1 All valves certificated as suitable for use as Type 3 valves must have undergone third-party testing to the requirements of Model Specification D08 and been awarded a licence under the TMV Scheme.

7.2 Each manufacturer claiming compliance will be able to provide purchasers with a copy of his acceptance letter and certificate of acceptance. This certificate will stand for a period of five years subject to successful auditing of the product.

7.3 The acceptance letter will include information to ensure the mixing valve can be readily identified and the designation codes under which the valve is acceptable can be confirmed.

If at any time a purchaser is in doubt as to the authenticity of any certificate or a product's acceptability, please contact the Secretary of the Technical Assessment Panel of the Thermostatic Mixing Valve Scheme for assistance (see Appendix II for address and contact number).

8.0 Designation and marking of Type 3 valves

Designation

8.1 TMVs complying with D08 are designated by:

a. the operating pressure range;

b. the intended application.

It is important that the valve is not used for any designations which the acceptance letter has not stipulated, as in these circumstances it cannot be regarded as a Type 3 valve.

The last element of the designation code is given in Table 4. In every case this element is preceded by "Thermostatic mixing valve – NHS Model Specification". Where a TMV is suitable for more than one application, the code can include the final element of each application, for example "Thermostatic mixing valve – NHS Model Specification -LP-BSW" would be suitable for bidet, shower or washbasin applications in the low pressure operating range. Where, for reasons of water economy, a valve has a flow rate of less than 8 *l*/min, it should carry the designation suffix 'E'.

Table 4 Designation codes

Code	Operating pressure range	Application
-HP-B	High pressure	Bidet
-HP-S	High pressure	Shower
-HP-W	High pressure	Washbasin
-HP-T44	High pressure	Bath with fill temperature up to 44°C
-HP-T46	High pressure	Bath with fill temperature up to 46°C
-LP-B	Low pressure	Bidet
-LP-S	Low pressure	Shower
-LP-W	Low pressure	Washbasin
-LP-T44	Low pressure	Bath with fill temperature up to 44°C
-LP-T46	Low pressure	Bath with fill temperature up to 46°C

High pressure = flow pressure hot and cold 1 to 5 bar
Low pressure = flow pressure hot and cold 0.2 to 1 bar

Marking

8.2 All Type 3 valves complying with the TMV Scheme will be permanently and legibly marked on the product with the manufacturer's name or identification mark and unique model reference, sufficient to enable the designation of the product to be compared with the corresponding information on any certificate of compliance with this Specification.

8.3 The certificate of acceptance will define the markings of each "listed" valve; any valve deviating from the unique model reference referred to in the acceptance letter should not be regarded as a Type 3 valve.

All Type 3 valves that meet the requirements of the Scheme will automatically have gained Water Byelaws Scheme Approval.

8.4 Where manufacturers make any changes to their marking this will be referred to officers of the Scheme who will issue a new acceptance letter which will update the changes.

9.0 Installation and operating instructions

9.1 For all types of valve the manufacturer should provide installation, operating and maintenance instructions.

9.2 For Type 3 valves the instructions should also include:

a. information on the designation of the thermostatic mixing valve concerned;

b. information on the commissioning and routine in-service tests to be performed;

c. information on the frequency of in-service tests and service work;

d. information on the need for any anti-backsiphonage devices (eg check valves) required to be installed with the mixing valve, together with the specification of such devices. This specification shall be sufficient to enable the combination of mixing valve and anti-backsiphonage devices tested in accordance with this specification to be replicated on site;

e. the need for the inclusion of any isolating valves etc to enable on-site tests to be made;

f. details of suitable outlet fittings (eg draw-off taps etc).

10.0 Surface temperature

Introduction

10.1 For most applications it is usual to design the low pressure hot water heating (LPHW) systems for 80°C flow or supply to the radiator with a 70°C return. Thus the surface temperature of the radiator, under normal design conditions, will vary between these values. Figure 1 shows that the average population will receive partial thickness burns in under a second at these temperatures and full thickness burns in about 10 seconds.

10.2 Certain sectors of the population, for example the elderly, mentally impaired and the young, will be at greater risk than the average population. Thus particular attention needs to be paid to the heating system temperatures.

Maximum safe temperatures

10.3 Space heating devices include thermal storage heaters, oil-filled radiators as well as conventional radiators, towel rails and other heating devices. Wherever patients, residents and visitors have access, the maximum surface temperature of space heating devices should not exceed 43°C when the system is running at the maximum design output.

10.4 Options to ensure safety are:

- guards should be fitted (see paragraph 10.7);
- low surface temperature heat emitters should be used;
- flow temperature reduction (temperature controls to fail to a safe position).

In Scotland, see SAN(SC)97/07 'Non-fire retardant polypropylene radiator covers: Fire risk and the English equivalent'. In England and Wales, see NHSE Hazard Notice (96)03, 'Non fire retardant plastic radiator covers'.

10.5 Choosing the first option will reduce the heat output by some 60%, hence more and/or larger radiators will be required. The preferable solution is the installation of purpose-manufactured low surface temperature heat emitters. Other design solutions which ensure that the maximum surface temperature does not exceed 43°C are acceptable.

Exposed pipework

10.6 Surface-mounted pipework which is exposed at low level, that is, within 2 m of the floor, presents an additional risk if it is carrying water above 43°C and should be securely insulated or "boxed" in. This includes vertical and horizontal pipe runs.

General

10.7 Where guards are used to prevent contact with hot surfaces, the size of the mesh should be carefully selected to prevent small hands penetrating and contacting the heat emitter. Guards should be cleaned regularly to prevent the build-up of debris.

10.8 The heat emitters should be designed so that:

- there are no surface hot spots;
- there are no access "holes" which may allow small hands to enter, that is, they should be totally enclosed;
- they are capable of being easily cleaned as a routine procedure (the cleaning procedure should be specified by the hospital infection control team).

References

Acts and Regulations

Health and Safety at Work etc Act 1974. The Stationery Office, 1974.

Nursing Homes and Nursing Agencies Act (Northern Ireland) 1971. The Stationery Office, 1971. (print on demand)

Registered Establishments (Scotland) Act 1987. The Stationery Office, 1987. (print on demand)

Registered Homes Act 1984. The Stationery Office 1984.

SI 1039: 1978 (NI 9) Health and Safety at Work (Northern Ireland) Order. The Stationery Office 1978.

SI 2051: 1992 The Management of Health and Safety at Work Regulations. The Stationery Office 1992.

SI 459: 1992 The Management of Health and Safety at Work Regulations (Northern Ireland). The Stationery Office 1992.

British Standards

BS 1415 Mixing valves
Part 1: 1976 Non-thermostatic, non-compensating mixing valves. BSI Standards, 1976.
Part 2: 1986 Specification for thermostatic mixing valves. BSI Standards, 1986.

BS 5779: 1979 (1987) Specification for spray mixing taps. BSI Standards, 1979.

BS 6700: 1987 Specification for design, installation, testing and maintenance of services supplying water for domestic use within buildings and their curtilages. BSI Standards, 1987.

Other publications

The control of legionellae in healthcare premises – a code of practice (Health Technical Memorandum 2040). NHS Estates, The Stationery Office 1994. (issued in 5 parts)

Controlling legionella in nursing homes and residential care homes (IND(G)253). Health and Safety Executive, HSE Books, 1997.

Health and safety in residential care homes (HS(G)104). Health and Safety Executive, HSE Books, 1993.

Hot and cold water supply, storage and mains services (Health Technical Memorandum 2027). NHS Estates, The Stationery Office 1995. (issued in 4 parts)

Model water byelaws (1986 edition). Department of the Environment, The Stationery Office, 1986.

Non-fire retardant polypropylene radiator covers: Fire risk and the English equivalent (SAN(SC)97/07). NHS in Scotland Management Executive, 1997.

Non fire-retardant plastic radiator covers (NHSE HN(96)03). NHS Estates, 1996.

Thermostatic mixing valves (healthcare premises) (Model Engineering Specification D08). NHS Estates, 1997.

Water fittings and materials directory. Water Research Centre, Richard Joseph Publishers, published annually.

Appendix I – Thermostatic mixing valves: sample maintenance log sheet

Hospital ______________________

Ward/location ______________________

Asset ident./Valve no ______________________

Installation date ______________________

Manufacturer ______________________

Model, type and designation of TMV ______________________

Point of delivery ______________________ see note 1

Set water temperature at outlet ______________________ °C see note 2

Location/access to TMV ______________________ see note 3

Location of hot and cold water isolation valves ______________________

Hot and cold water supply pressure ______________________

History see note 4

Date	Work details	Signature of skilled or responsible person

Notes

1. Point of delivery: bath, shower, washhand basin, bidet, etc
2. Set water temp at outlet: refer to Trust policy guidance and risk assessments
3. Location/access of/to the TMV – under washhand basin, panel access in wall, ceiling or floor etc
4. History: state whether new, new replacement, refurbished, what parts were replaced, adjustment made etc. All entries must be signed by the responsible or skilled person

When any work or adjustments are made to the TMV, always record the TMV water outlet temperature BEFORE and AFTER the work or adjustment to the TMV.

Appendix II: Contact names and addresses

For England and Wales:

NHS Estates
1 Trevelyan Square
Boar Lane
LEEDS LS1 6AE
Tel: 0113 254 7000

For Scotland:

The NHS in Scotland
Healthcare Engineering and Environment Unit
The University of Strathclyde
Room 8.51 Graham Hills Building
50 George Street
GLASGOW G1 1QE
Tel: 0141 548 3446

Hazard Co-ordinator
Scottish Healthcare Supplies
Trinity Park House
South Trinity Road
EDINBURGH EH4 3UE
Tel: 0131 551 8402

For Northern Ireland:

Health Estates, Estate Policy
Stoney Road
Dundonald
BELFAST BT16 0US
Tel: 01232 520025

Health and Safety Executive

375 West George Street
GLASGOW G2 4LW
Tel: 0141 275 3000

Belford House
59 Belford Road
EDINBURGH EH4 3UE
Tel: 0131 247 2000

14 Cardiff Road
Luton
BEDFORDSHIRE IU1 1PP
Tel: 0158 244 4200

Water Research Council

WRc Evaluation and Testing Centre
Fern Close
Pen-y-Fan Industrial Estate
Oakdale
GWENT NP1 4EH
Tel: 01495 248454

Other publications in this series

The Pressure Systems and Transportable Gas Containers Regulations 1989, ISBN 0-11-321674-2, price £15

Clinical waste incineration – Joint venture applications, ISBN 0-11-321768-4, price £15

Safe disposal of clinical waste – Whole hospital policy guidance, ISBN 0-11-322196-7, price £15

High voltage (safety) training, ISBN 0-11-322225-4, price £30

Structured cabling for IT systems, ISBN 0-11-322229-7, price £30

Static discharges, ISBN 0-11-322039-1, price £30

Magnetic resonance imaging, ISBN 0-11-322051-0, price £95

Telemedicine, ISBN 0-11-322061-8, price £60

Clinical waste in the community (in preparation)

Water supply and waste water disposal (in preparation)

About NHS Estates

NHS Estates is an Executive Agency of the Department of Health and is involved with all aspects of health estate management, development and maintenance. The Agency has a dynamic fund of knowledge which it has acquired during over 30 years of working in the field. Using this knowledge NHS Estates has developed products which are unique in range and depth. These are described below. NHS Estates also makes its experience available to the field through its consultancy services.

Enquiries about NHS Estates should be addressed to:
NHS Estates, Publications Unit, Department of Health,
1 Trevelyan Square, Boar Lane, Leeds LS1 6AE.
Telephone 0113 254 7000.
http://www.demon.co.uk/nhsestates/hpage.html/

Some NHS Estates products

Activity DataBase – a computerised briefing and design system for use in health buildings, applicable to both new build and refurbishment schemes. *NHS Estates*

Design Guides – complementary to Health Building Notes, Design Guides provide advice for planners and designers about subjects not appropriate to the Health Building Notes series. *SO*

Estatecode – user manual for managing a health estate. Includes a recommended methodology for property appraisal and provides a basis for integration of the estate into corporate business planning. *SO*

Concode – outlines proven methods of selecting contracts and commissioning consultants. Reflects official policy on contract procedures. *SO*

Works Information Management System – a computerised information system for estate management tasks, enabling tangible assets to be put into the context of servicing requirements. *NHS Estates*

Health Building Notes – advice for project teams procuring new buildings and adapting or extending existing buildings. *SO*

Health Guidance Notes – an occasional series of publications which respond to changes in Department of Health policy or reflect changing NHS operational management. Each deals with a specific topic and is complementary to a related HTM. *SO*

Health Technical Memoranda – guidance on the design, installation and running of specialised building service systems, and on specialised building components. *SO*

Health Facilities Notes – debate current and topical issues of concern across all areas of healthcare provision. *SO*

Encode – shows how to plan and implement a policy of energy efficiency in a building. *SO*

Firecode – for policy, technical guidance and specialist aspects of fire precautions. *SO*

Capital Investment Manual Database – software support for managing the capital programme. Compatible with Capital Investment Manual. *NHS Estates*

Model Engineering Specifications – comprehensive advice used in briefing consultants, contractors and suppliers of healthcare engineering services to meet Departmental policy and best practice guidance. *NHS Estates*

Quarterly Briefing – gives a regular overview on the construction industry and an outlook on how this may affect building projects in the health sector, in particular the impact on business prices. Also provides information on new and revised cost allowances for health buildings. Published four times a year; available on subscription direct from NHS Estates. *NHS Estates*

Items noted "SO" can be purchased from The Stationery Office Bookshops in London (post orders to PO Box 276, SW8 5DT), Edinburgh, Belfast, Manchester, Birmingham and Bristol or through good booksellers.

NHS Estates consultancy service

Designed to meet a range of needs from advice on the oversight of estates management functions to a much fuller collaboration for particularly innovative or exemplary projects.

Enquiries should be addressed to: NHS Estates Consultancy Service (address as above).

Printed in the United Kingdom for The Stationery Office
J0042514 C10 3/98 9385/7774

NHS Estates
An Executive Agency of the Department of Health

Department of Health
1 Trevelyan Square
Boar Lane
Leeds
LS1 6AE
Fax 0113 254 7299
Telephone 0113 254 7000

Dear Customer

FOLD

At NHS Estates, we constantly strive to produce publications that are relevant to the demands of the NHS and our worldwide healthcare customers.

It would help us to provide you with a better service if you could set aside a few minutes to complete the questionnaire on the reverse, fold as shown and return in a window envelope.

Thank you for your comments.

Kate Priestley

Chief Executive

FOLD

KATE PRIESTLEY
CHIEF EXECUTIVE
NHS ESTATES
FREEPOST LS 5588
LEEDS
LS1 1YY

March 1998 – HGN 'Safe' hot water and surface temperatures

Publication title .... Date ....

Reader details

Name ....

Position ....

Organisation ....

Address ....

.... Postcode ....

Telephone ....

Please tick the appropriate box	Very Good	Good	Fair	Poor
Appropriateness of subject	❑	❑	❑	❑
Met my need	❑	❑	❑	❑
Clarity of presentation	❑	❑	❑	❑
Quality of content	❑	❑	❑	❑
Value for money	❑	❑	❑	❑
Overall rating of the publication	❑	❑	❑	❑

Will you be purchasing future editions in the series? Yes ❑ No ❑

Other comments/areas for improvement?

....

....

Do you have any suggested subjects for the future?

....

....

Do you receive copies of our Newslink newsletter? YES ❑ NO ❑

Do you wish to receive a copy? YES ❑ NO ❑

Have you purchased other products/services from us recently?

Please tick

Other publications ❑ **Activity DataBase** ❑

Consultancy services ❑ **Wims software enhancements** ❑

Other software products ❑ **Other: please specify** ❑

Are there other ways in which NHS Estates can assist you?

....

....

....

How did you receive your copy of this document?

Sent/issued by NHSE ❑ **Purchased from The Stationery Office** ❑